This book belongs to

....

and Sue

Notes for Grown-ups

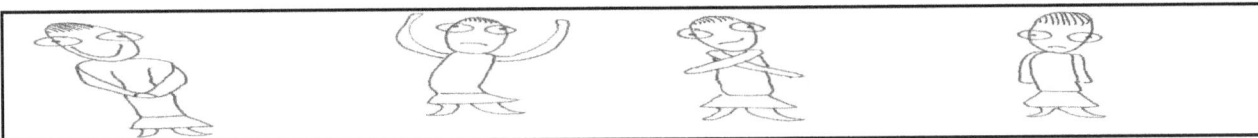

Comic Maths: Sue (Key Stage 1, Level 1) has been created for children by reporting on children's own mathematical language, ideas and reasoning. It supports fantasy-based learning by using story-lines, comic characters and crazy situations in an attempt to embed mathematics into children's' everyday lives. This book is organized as a series of 10 short comics with the following National Curriculum focus.

Mathematics Key Stage 1 Level 1					
Comic	Processes in Mathematics	Number	Measures	Shape and Space	Handling Data
1	●	●			
2	●	●			
3	●	●			
4	●	●			
5	●	●			
6	●	●			
7	●		●		
8	●		●		
9	●			●	
10	●				●

There are extension activities corresponding to each comic in 'Extra Sums for Greedy People'. 'Crazy *Baby*' pages aim to get us to gaze out of the widow and think crazy thoughts! Story Pages aim to lead the reader through the book. 'Answers in The Back' gives answers to the questions set in each comic, the extension activities, and to the 'Crazy *Baby*' thoughts.

Comic Maths

Sue

Fantasy-based learning for 4, 5 and 6 year olds

Brian Williamson

Level 1

Cover design by
Kathryn Wilson

Six ideas for using this book:

1. Act out Sue, John and Anne travelling though their comic adventures! Try to bring the playground into the classroom! Make costumes for the Comic Maths characters and put on a Comic Maths Show.

2. The drawings in this book are line-drawings so use Comic Maths as a colouring book. Visit the story pages that link the comics, colour them in and then find your way into a comic or two!

3. Use Comic Maths as a learning support resource. Look up those areas of the curriculum recently covered at school and see if Comic Maths can help.

4. Use Comic Maths as a fantasy-based home-schooling resource.

5. Show your young friend the exercises in 'Extra Sums for Greedy People!' observe their response and use this information to assess their learning needs.

6. Just leave it around and see what happens!

All pages marked: **COPY ME** © Brian Williamson 2011 Comic Maths may be photocopied but not for commercial gain.

Welcome to Comic Maths!

Sue, John and Anne, Betty, Charlie the Monkey, Bill, Granddad, Anne's Mum, Kilo and Milli.

Meet Sue

Sue is a good friend.

She helps John.   She helps Anne.

She has a fight with a cardboard box!

She helps Charlie the Monkey count footballs.

She NEVER EVER takes her party hat off ... and hedgehog wants her ice cream!

The Chooser

Comic 1	1 2 3	11
Comic 2	Colour Count.	17
Comic 3	Football Pictures.	26
Comic 4	Play Write.	34
Comic 5	Charlie and the Monkey.	40
Comic 6	What Comes Next?	45
Comic 7	My Square is Bigger than your Square.	50
Comic 8	Bedtime.	55
Comic 9	Cardboard Castles.	65
Comic 10	Silly People.	71
	Extra sums for greedy people.	75
	Answers in the Back.	90

Story Page

COMIC MATHS

1 2 3

colour in

CRAZY BABY! thinks about 3

Three

It is a word?

It is a number.

One two three 1, 2, 3

3

3 elephants

3

3 suns

3 hiccups

3 fish. 3 chips. 3 chairs. 3 cats. 3 apples. 3 grandmas on toast. 3 friends. 3 houses. 3 trees. 3 flowers. 3 bowls of rice pudding.

Say more!!

Story Page

COMIC MATHS

Colour Count 2

John drew some animals.

Sue coloured them in and counted.

There are four animals.

John drew some more animals.
Now you colour and count.

1

2

There are ___ animals.

John drew some aliens. Colour count these.

There are ___ aliens.

Thanks for telling me!

Colour and count the Comic Maths Family!

There are _____ of them.

COPY ME!
© Brian Williamson 2011
Comic Maths

 CRAZY BABY! counts the aliens.

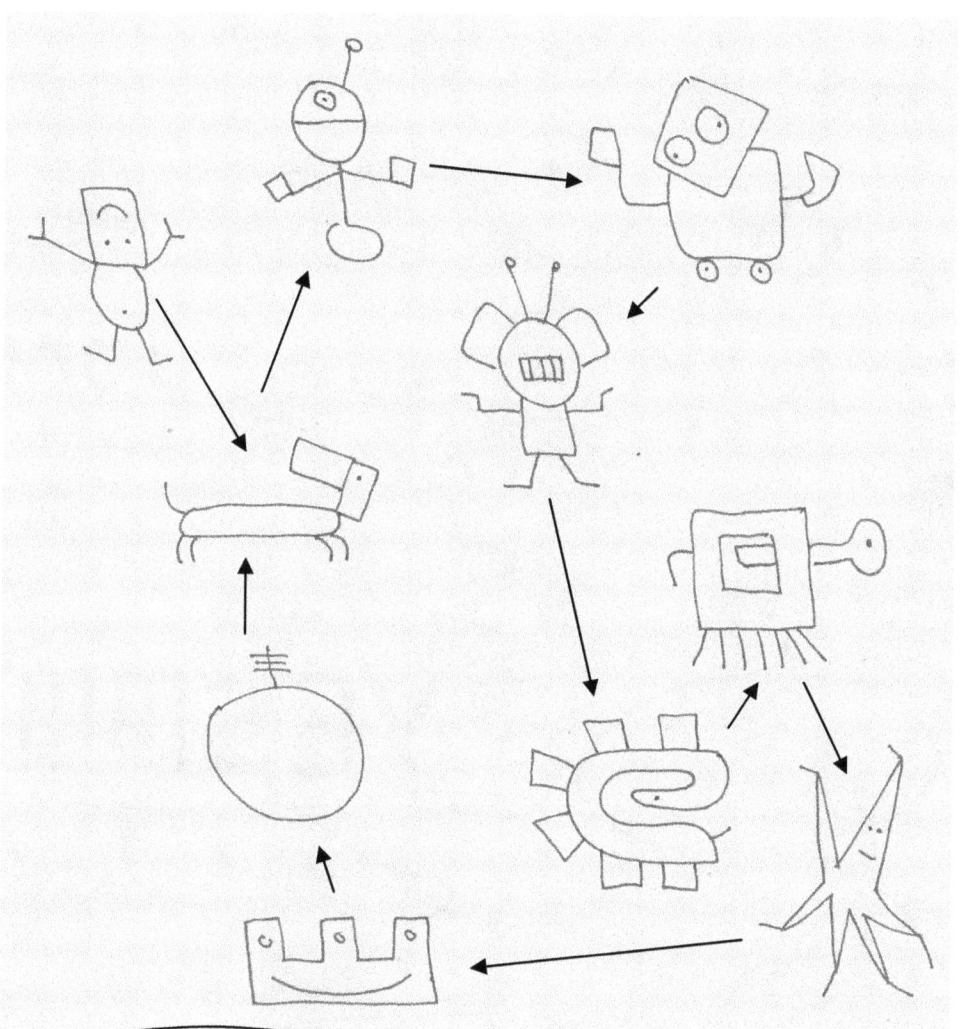

Which way did you count?

My way CRAZY BABY!

Story Page

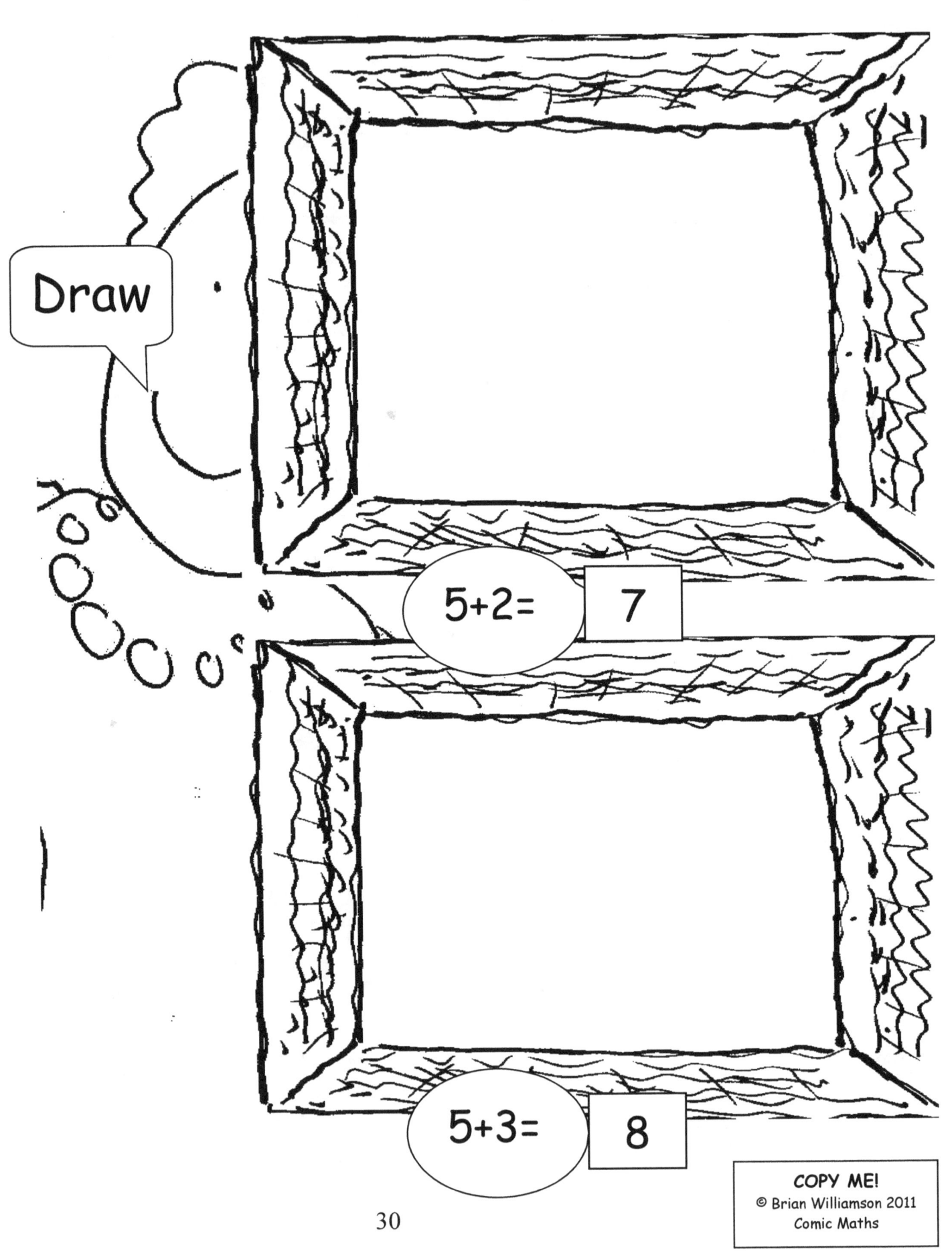

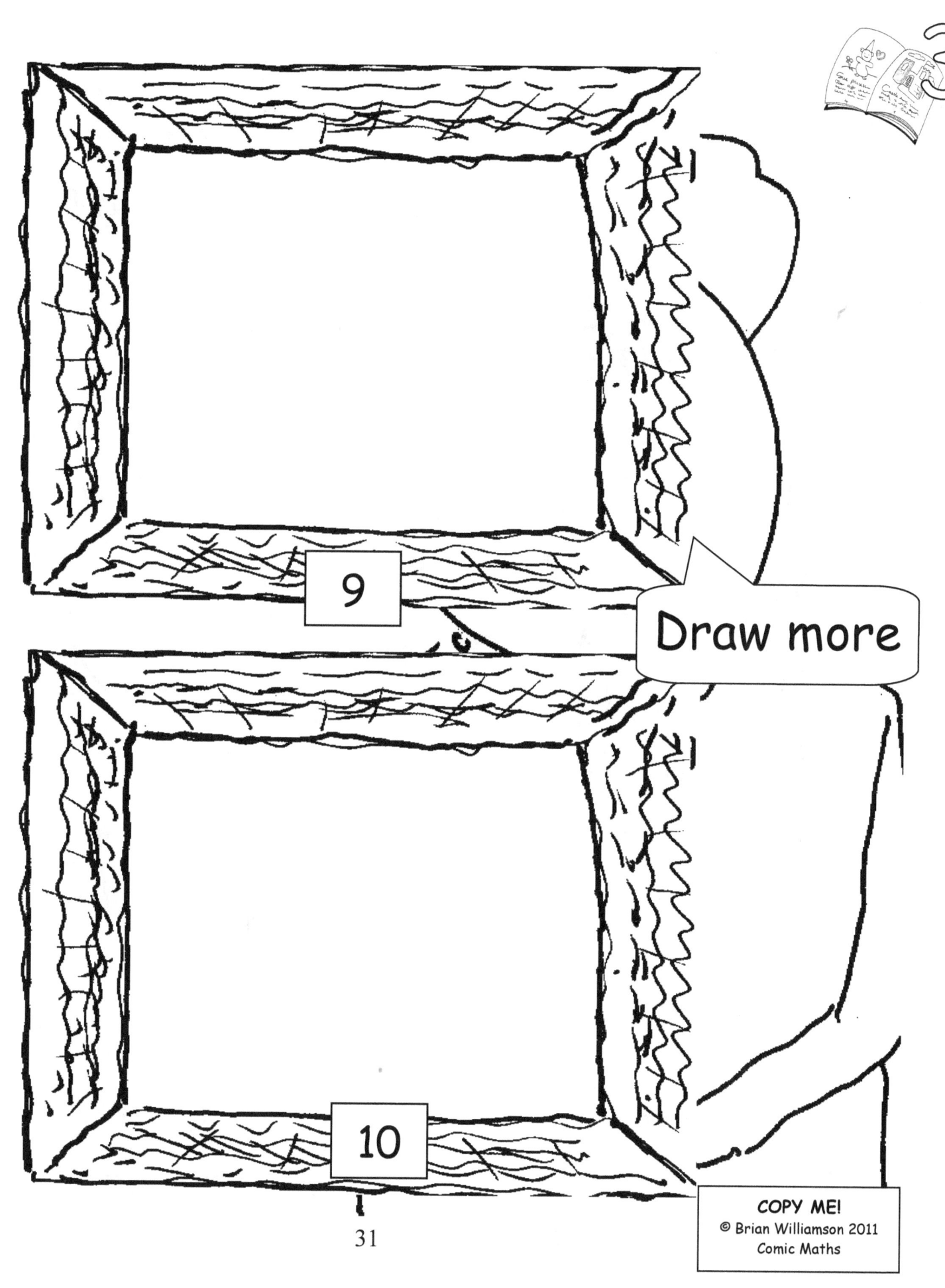

CRAZY BABY! thinks 7 and 9

We grow the same way.

I am like a flower!

What do you think 8 is like?
8 is like a _____

Story Page

Yes. Look.

 1

 2

Make a face for 3.

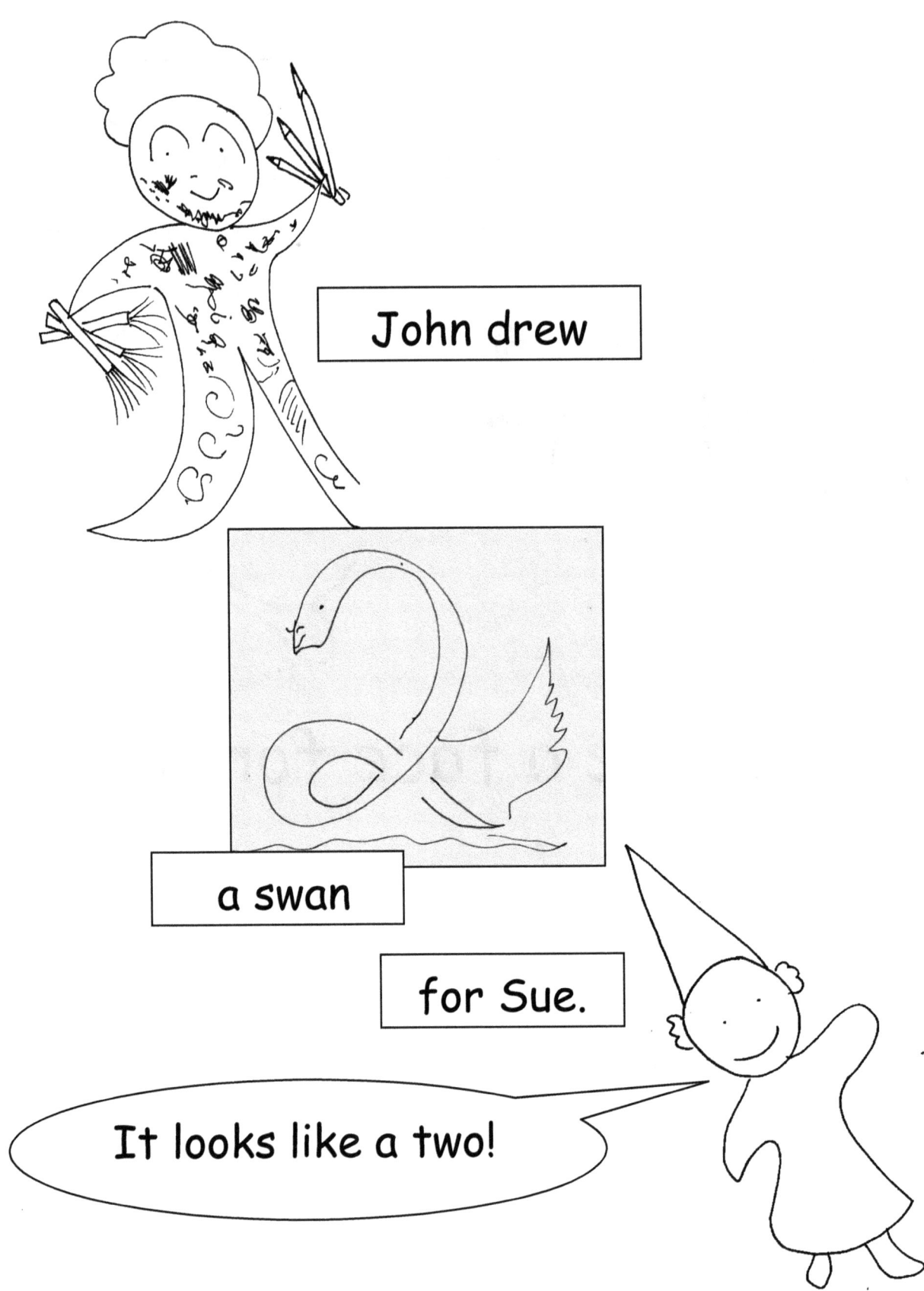

John is a SPACE ROCKET holding a pen!
Making rocket noises!! Making an eight!!

Eights are GREAT!!

CRAZY BABY! likes this picture.

A beautiful picture!

Is he a **CRAZY BABY?**

_____

Story Page

Sue John and Anne meet Charlie the Monkey.

COMIC MATHS 5

Charlie the Monkey the Mad Guesser!!!

CRAZY BABY! learns =

 =

5 = 5

Sue knows this!

Story Page

John, Sue and Anne meet Betty the fortune teller.

COMIC MATHS 6

WHAT COMES NEXT?????

I wish I knew.

I wish I knew.

Draw the next monster . . .

 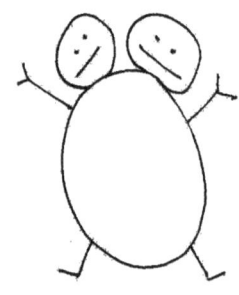 

Draw the next hedgehog . . .

 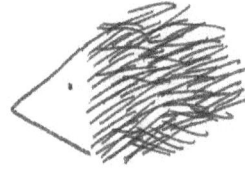 

Draw the next fluffy cloud sheep . . .

Draw the next Anne's face . . .

COPY ME!
© Brian Williamson 2011
Comic Maths

Mess or simple? ....

Story Page

COMIC MATHS

My square is bigger than your square!

Colour in the person with the longest arms.

Colour in the heaviest person.

Colour in the person with the biggest drink.

COPY ME!
© Brian Williamson 2011
Comic Maths

Colour in the person with the biggest square!

Play this game with your friends.
  Who can find the biggest ball?
  Who can find the biggest spoon?
  Who is the tallest?
  Who has the longest nose?

COPY ME!
© Brian Williamson 2011
Comic Maths

CRAZY BABY!
meets monster number

Monster numbers need a lot of room. Don't they?

What is the biggest number you know?
_____

COPY ME!
© Brian Williamson 2011
Comic Maths

Story Page

Most days John . . .

 Eats his breakfast at 7 o'clock in the morning!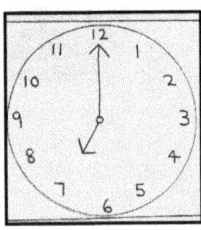

 Cleans his teeth at 8 o'clock.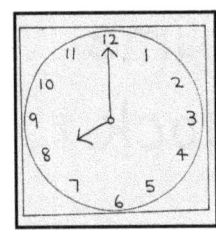

 Plays maths with his Teacher at 10 o'clock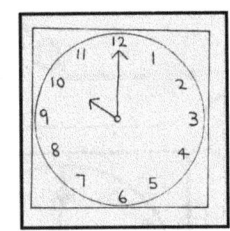

 Eats his lunch at 12 o'clock NOON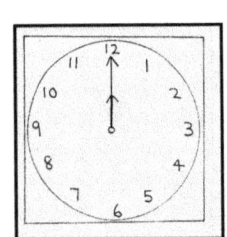

 Plays outside at 3 o'clock.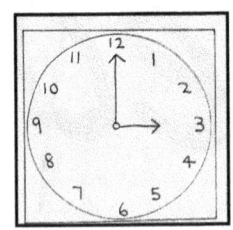

 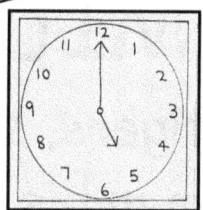 

"Don't like sausages."

Eats his tea at 5 o'clock.

 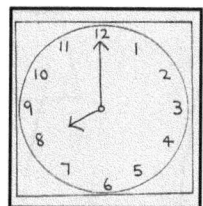

Cleans his teeth at 6 o'clock.

 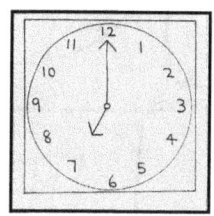

Reads a comic about castles and Princess Sue at 7 o'clock.

and at 8 o'clock it is ...

Bedtime

and...

at 12 o'clock MIDNIGHT John dreams of scratchy tigers, smiling angels and being at sea in a storm in his bed!

Make the clocks show the times John ...

Did you know?
One day John ...

played maths with his teacher at MIDNIGHT!

... and went to bed at 3 o'clock in the afternoon.

Change John's day to make it as CRAZY as you like ...

Teeth at   o'clock and   o'clock.

 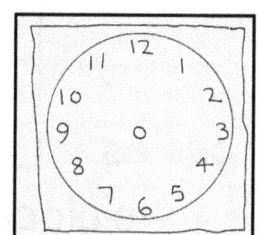

Breakfast at   o'clock.

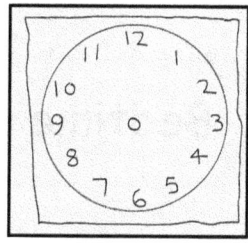 

Play maths at   o'clock.

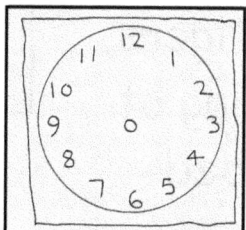

Lunch at   o'clock.

COPY ME!
© Brian Williamson 2011
Comic Maths

Plays outside at   o'clock.

Eats his tea at   o'clock.

Reads a story about castles and Princess Sue at   o'clock.

Bedtime at   o'clock.

Dreams of scratchy tigers, smiling angels and being at sea in a storm in his bed at   o'clock.

COPY ME!
© Brian Williamson 2011
Comic Maths

CRAZY BABY! thinks it's always Sunday

7 days after Sunday is Sunday.
14 days after Sunday is Sunday.
21 days after Sunday is Sunday.
28 days after Sunday is Sunday.
35 days after Sunday is Sunday.

BILL, why is it ALWAYS Sunday?

Story Page

John, Sue, Anne, Anne's Mum, Granddad and Milli are going on holiday ...

... to a beech!!!!!

COMIC MATHS 9
CARDBOARD CASTLES

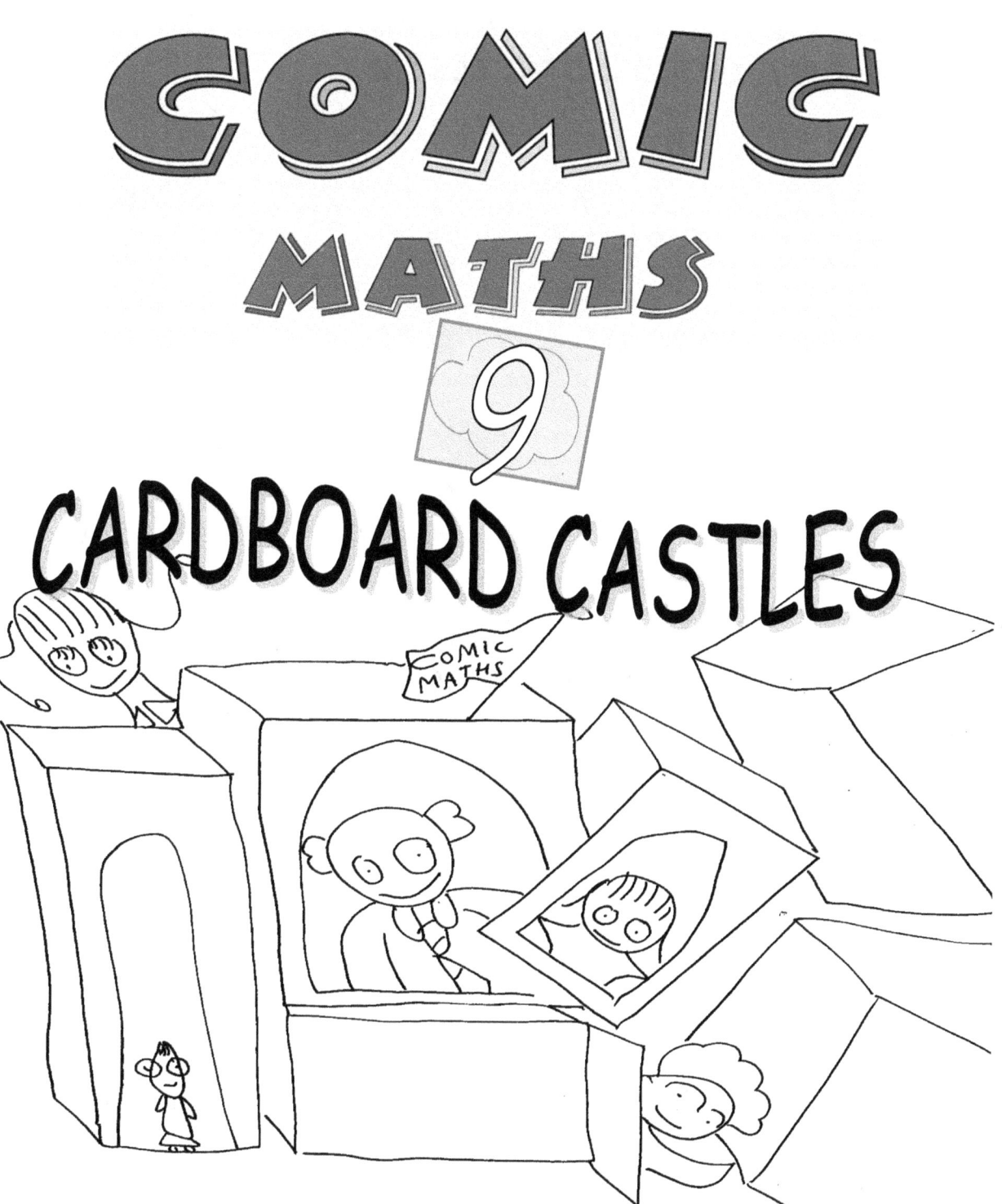

John, Anne, Mili and Granddad made a castle using cardboard boxes. They asked Anne's Mum to cut out holes for the windows.

Sue had a fight with a cardboard box.
It nearly knocked her hat off!!

"I will fight you box!!"

"Under"

"Over"

"Inside"

"Behind"

"AND in front!!"

You try making a castle out of cardboard boxes! Ask a grown-up to cut the windows out and **enjoy!**

CRAZY BABY!

meets a shape family

Mr. and Mrs. Square and their children . . .

.... Rectangle and Triangle.

Do you know someone who looks like a triangle?

Story Page

COMIC MATHS 10
Silly People!!!

Some people are silly.
Some people are not.

Who do you think is silly in these pictures?
Colour their faces **red** and count them.

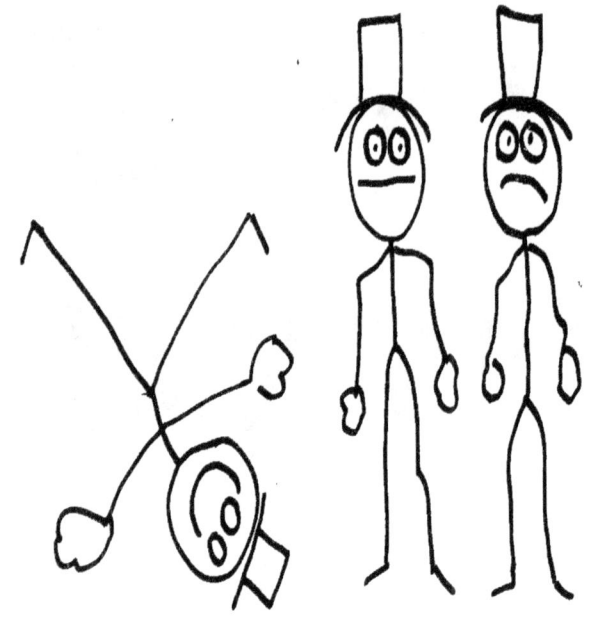

How many?

Silly	
Not Silly	
Everyone	

How many?

Silly	
Not Silly	
Everyone	

COPY ME!
© Brian Williamson 2011
Comic Maths

How many?

Silly	
Not Silly	
Everyone	

Tell a friend how you knew
that a person was silly.

COPY ME!
© Brian Williamson 2011
Comic Maths

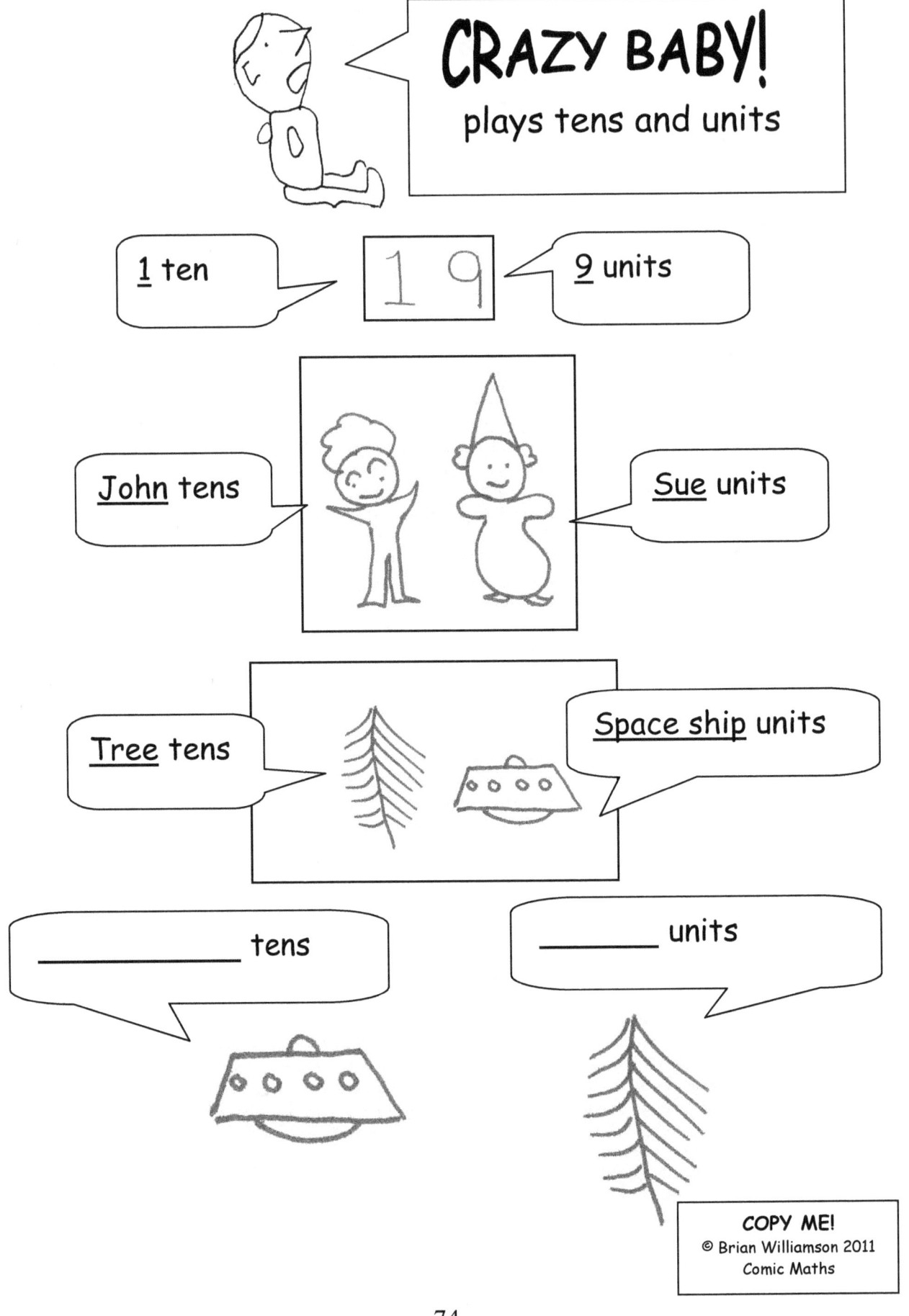

Extra Sums for Greedy People!

COMIC MATHS

1 2 3

Find a piece of paper.
Find a pencil.

Draw crazy baby. 4 hedgehogs
4 flowers. 4 petals. 4 leaves.
4 buzzy bees. 4 fluffy clouds.
Colour in!

COMIC MATHS
Colour Count 2

There are .... pieces.

There are .... pieces.

There are .... pieces.

There are ……… pieces.

COMIC MATHS Football Pictures 3

Draw football pictures for 15, 20 and 25.

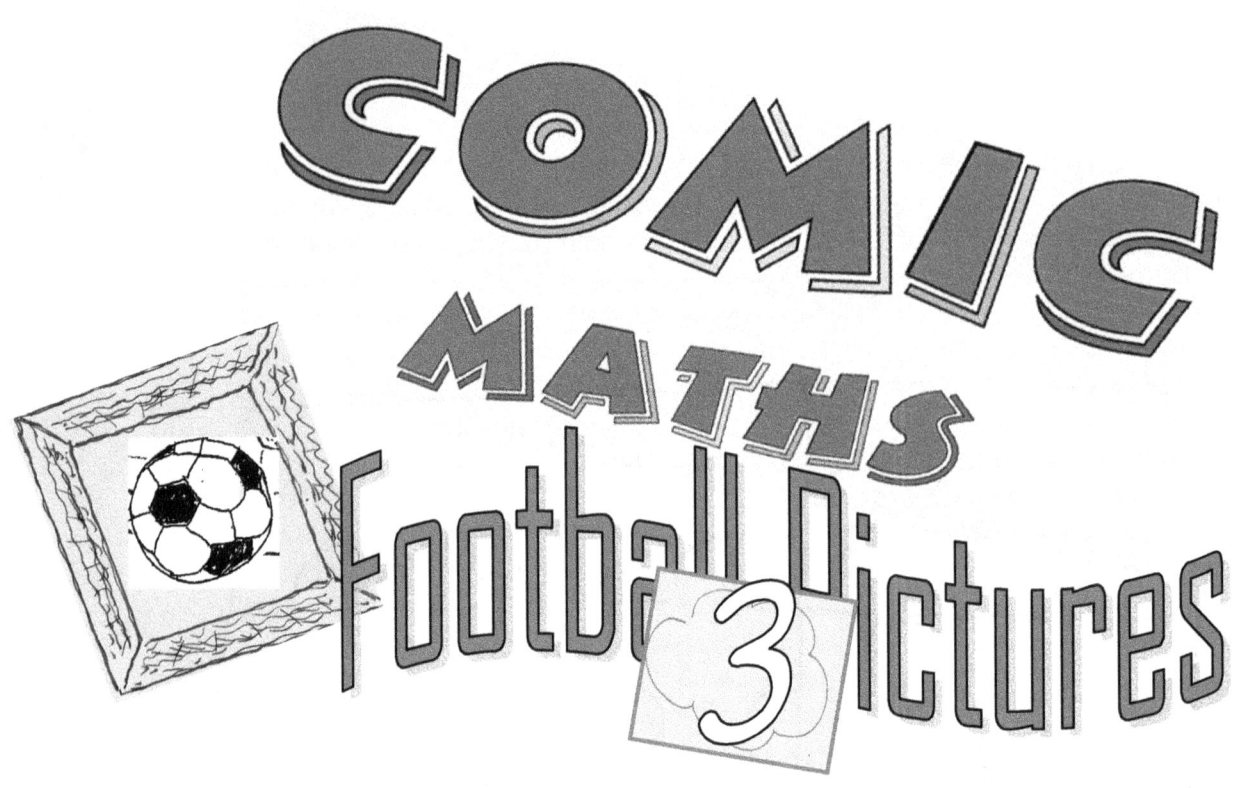

5+5+5

15

COPY ME!
© Brian Williamson 2011
Comic Maths

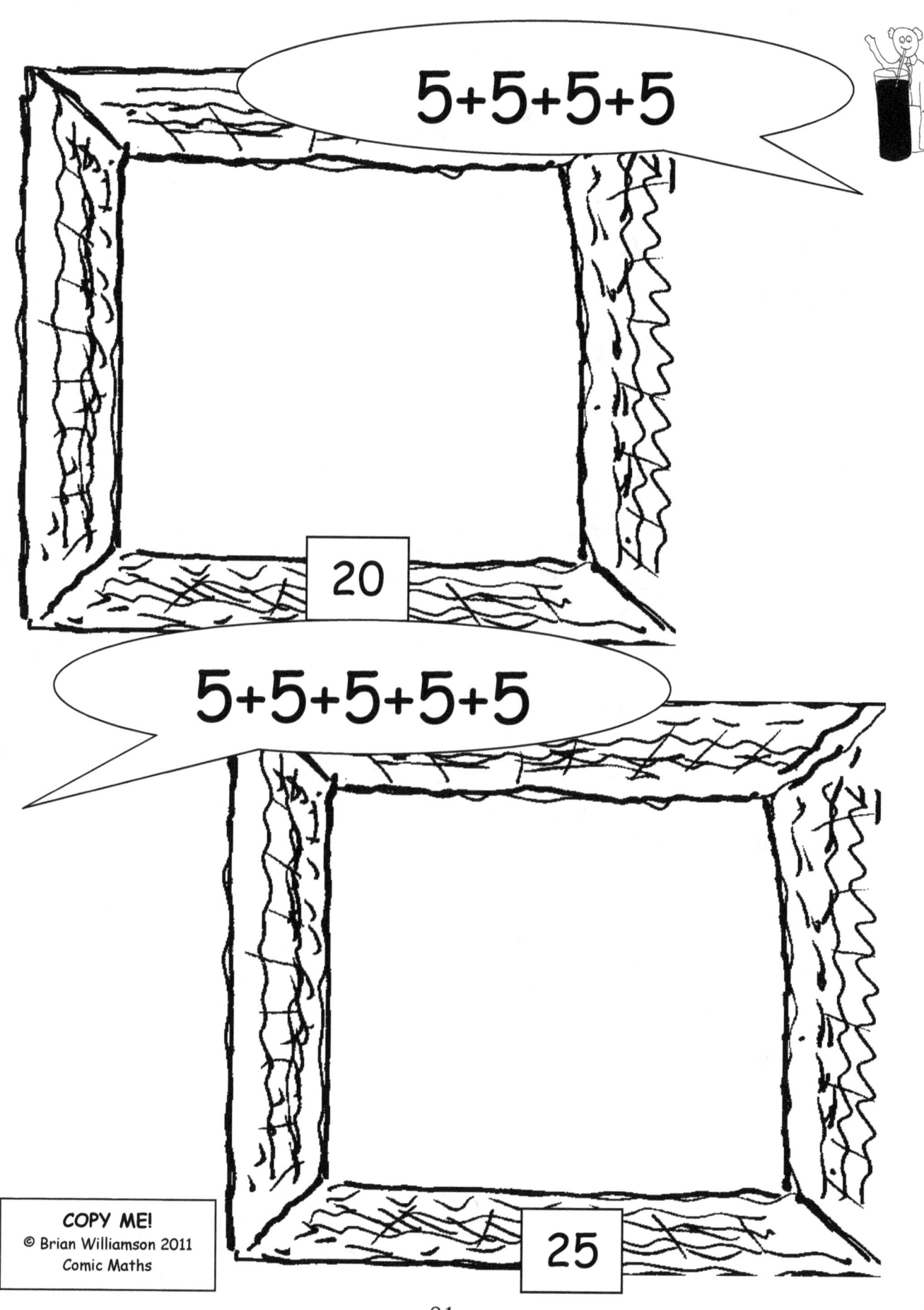

Close your eyes really tight!
Think of footballs... one

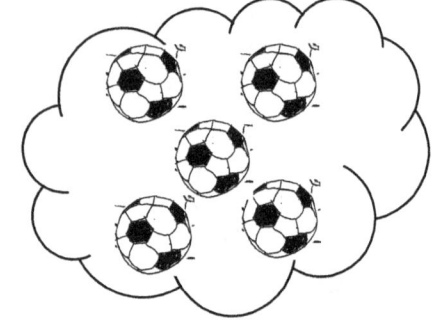

... five ...

... ten ...

... any!

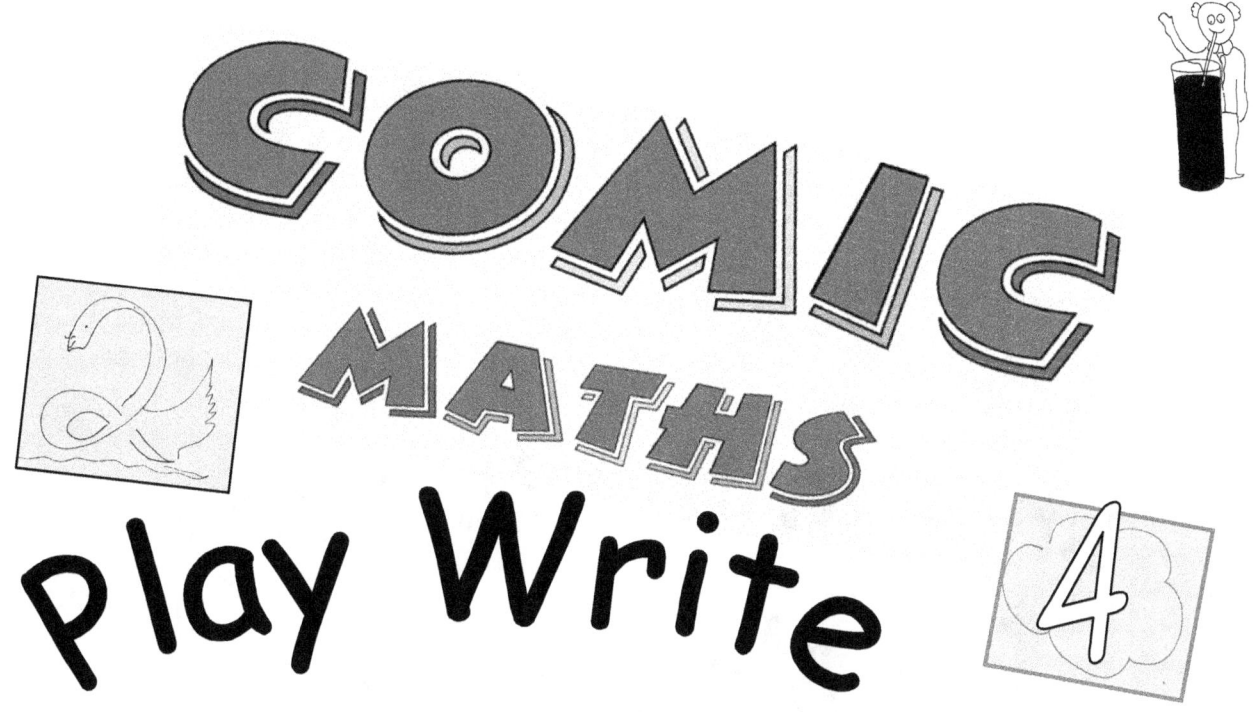

Play Write

Draw a face made out of the numbers 0, 0, 1, 3, 5 and 7.

Copy this swan and make a two!

COPY ME!
© Brian Williamson 2011
Comic Maths

COMIC MATHS 5

Charlie the Monkey the Mad Guesser!!!

Be Charlie and surprise a friend!!

Look
There are 66 fingers and 14 thumbs on my hand!

Look
There are ___ windows in that house

CRAZY NUMBERS!

Look
You have ___ noses on your face and you are wearing ___ pairs of trousers.

COPY ME!
© Brian Williamson 2011
Comic Maths

wHAT COMES nEXT?????

A B C D E ___

A C E G I ___

2 4 6 8 10 ___

5 10 15 20 25 ___

6 11 16 21 26 ___

10 20 30 40 50 ___

100 90 80 70 60 ___

M T W T F ___

J F M A M ___

1 2 3 1 2 3 1 2 ___

COPY ME!
© Brian Williamson 2011
Comic Maths

COMIC MATHS

My nose is bigger than your nose!

Line up.
Big noses first.

Line up.
Big feet first.

COMIC MATHS 8

Bedtime

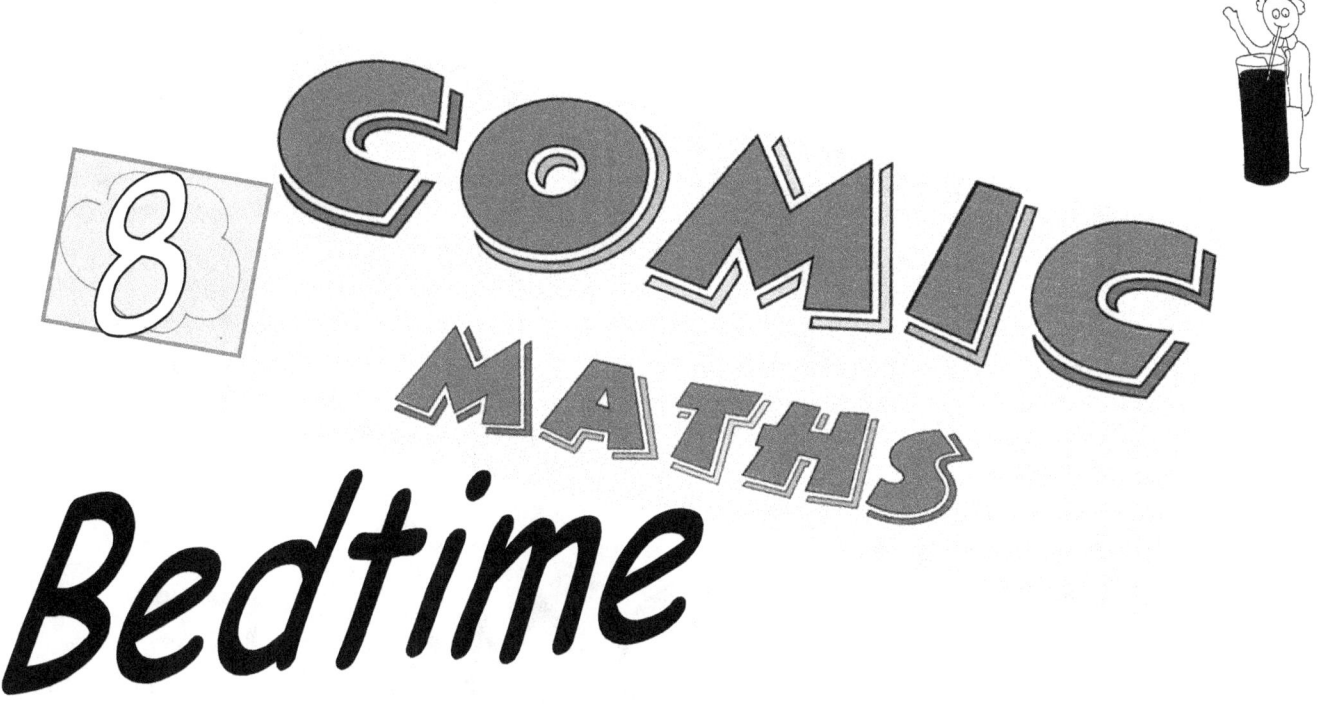

Wear a watch.

Ask the time at
1 o'clock,
2 o'clock,
3 o'clock,
4 o'clock,
5 o'clock,
6 o'clock,
7 o'clock,
8 o'clock,
9 o'clock,
and 10 o'clock.

Just checking!!!!

COMIC MATHS 9
CARDBOARD CASTLES

Have a fight
with a cardboard box.
Who won?

_____

Make things.
Talk about it.

The tooth brush
is inside.
The tin
is on top.
The ball
is at the front.

This is my art.

BE SAFE

Ask a grown-up

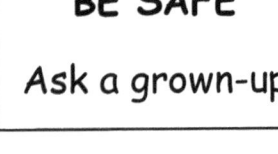

My Art

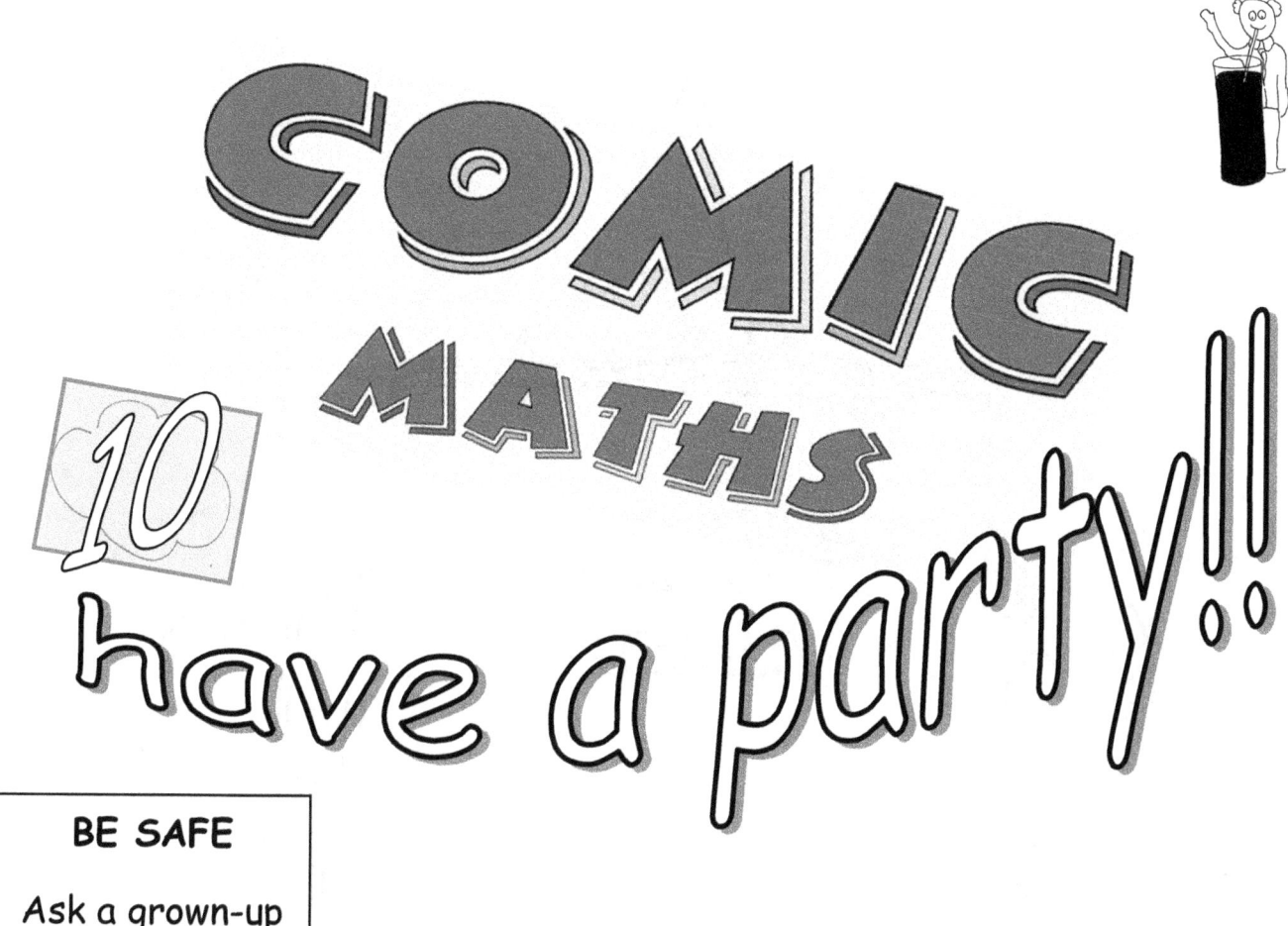

COMIC MATHS 10
have a party!!

BE SAFE

Ask a grown-up

Tell anyone who is silly to stand in the corner.

How many? _____

Tell the silly boys to say "boy silly".

How many? _____

Tell the girls to jump up and down.

How many? _____

Tell anyone who has a head to put their hand up.

How many? _____

Answers in The Back

COMIC MATHS 1

Your best colouring, is the answer!

CRAZY BABY!

said 3 clocks. 3 birthdays. 3 onions.

COMIC MATHS
Colour Count

There are <u>6</u> animals.
There are <u>10</u> aliens.
There are <u>10</u> of them.

CRAZY BABY!

said your way is the answer.

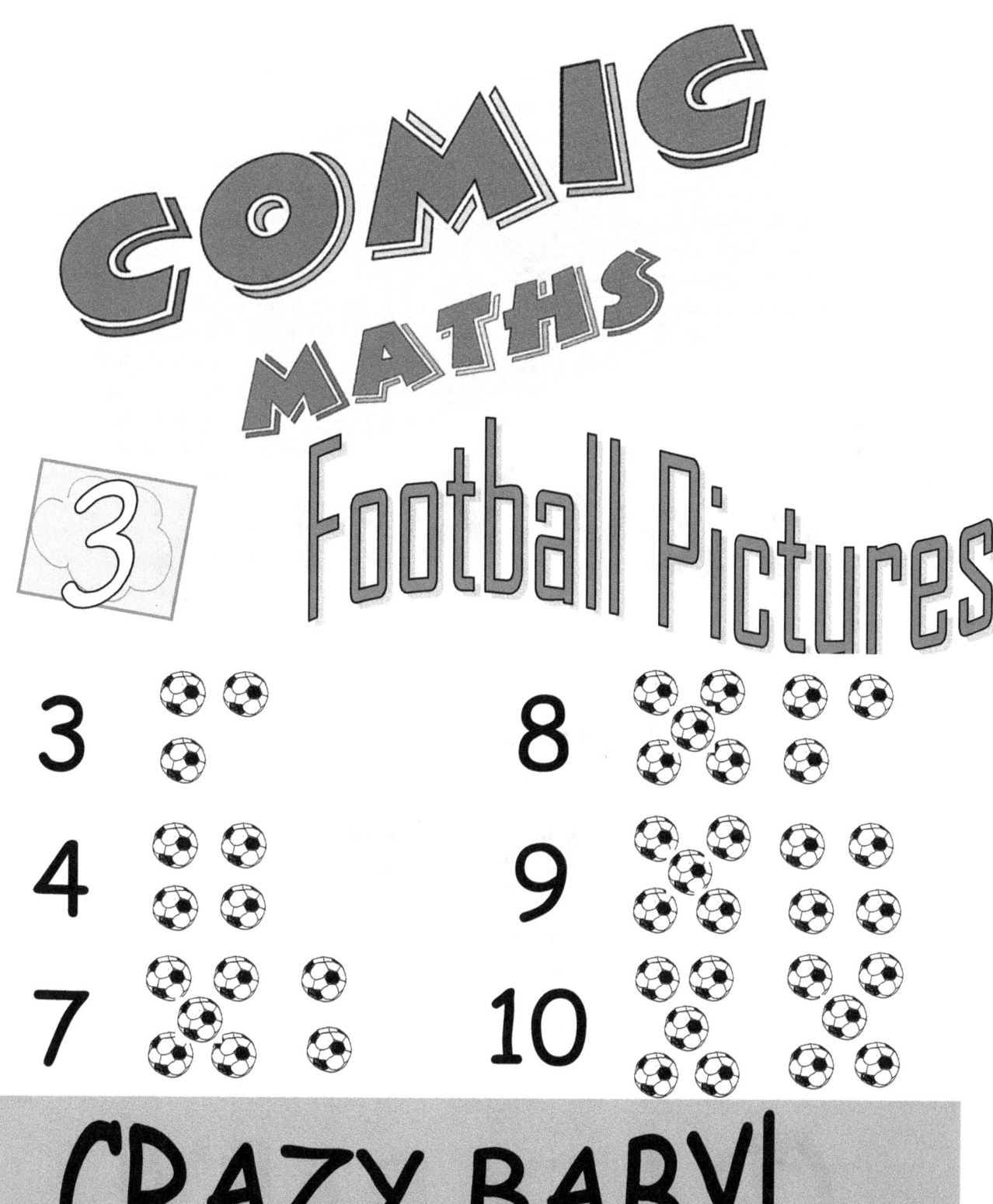

COMIC MATHS 4

Play Write

Very noisy.
Very fast.
Very big.

CRAZY BABY!

said no. All numbers are beautiful.

COMIC MATHS 5

Charlie the Monkey the Mad Guesser!!!

Not 9.

Not 1.

CRAZY BABY!

said

a chair is a chair.
an elephant is an elephant.
five is five.
y is 3x + 4

WHAT COMES NEXT?????

CRAZY BABY!

said mess and simple.

COMIC MATHS 7

My square is bigger than your square!

Colour in
John
Kilo
Granddad

CRAZY BABY!
said
54676766799999395593
63868385868 4483965896

My ears are bigger than your ears!

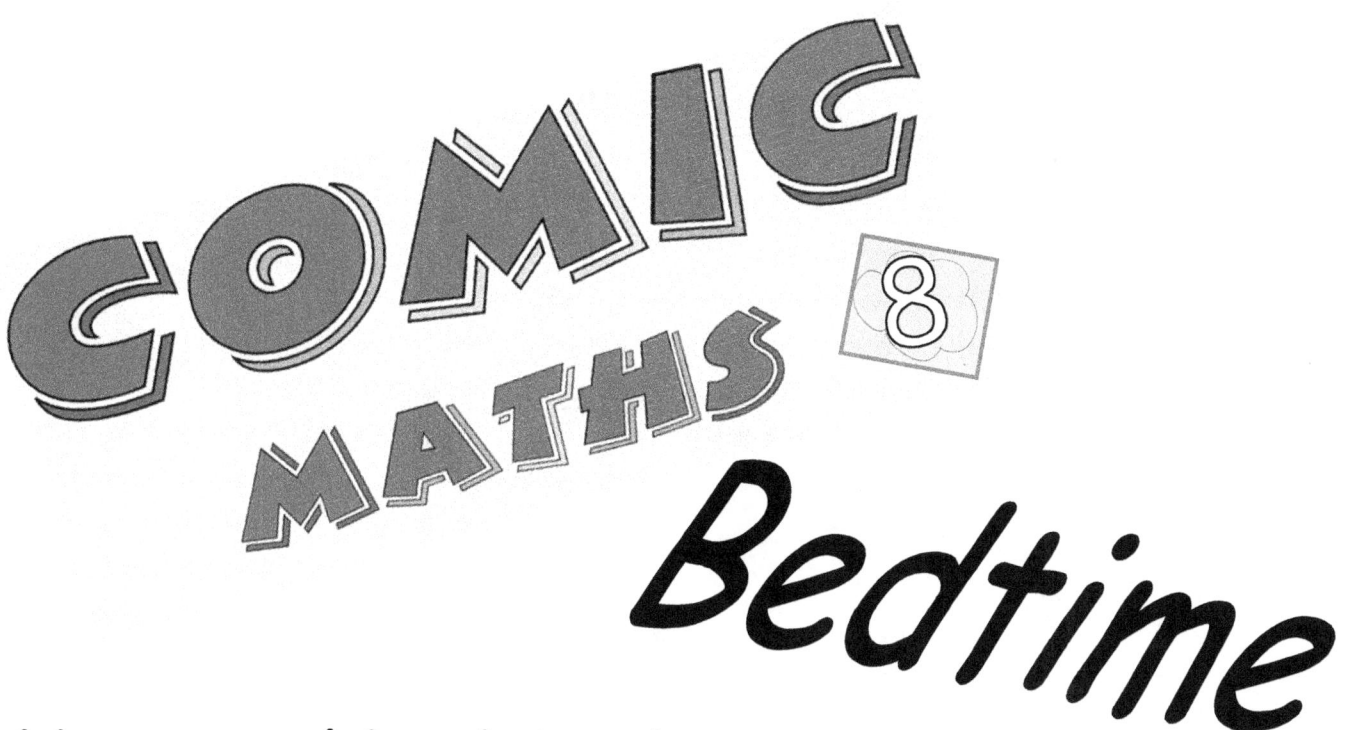

Bedtime

Your real bedtime!

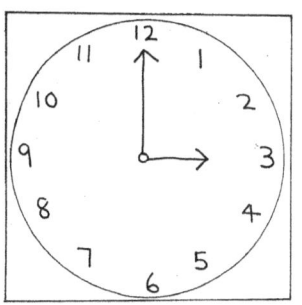

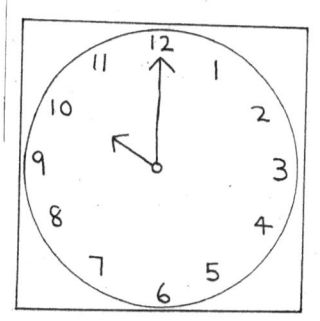

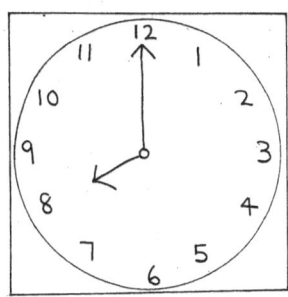

CRAZY BABY!
said because there are 7 days in a week.

COMIC MATHS 9

CARDBOARD CASTLES

Draw the very best castle you can!
Use the words
<u>under</u>, <u>over</u>, <u>inside</u>, <u>behind</u>

CRAZY BABY!

said Sue's hat looks like a triangle.

COMIC MATHS 10

Silly People!!!

Silly	1
Not Silly	2
Everyone	3

Silly	2
Not Silly	3
Everyone	5

Silly	6
Not Silly	6
Everyone	11

CRAZY BABY!

said   <u>Space ship</u> tens   <u>Tree</u> units

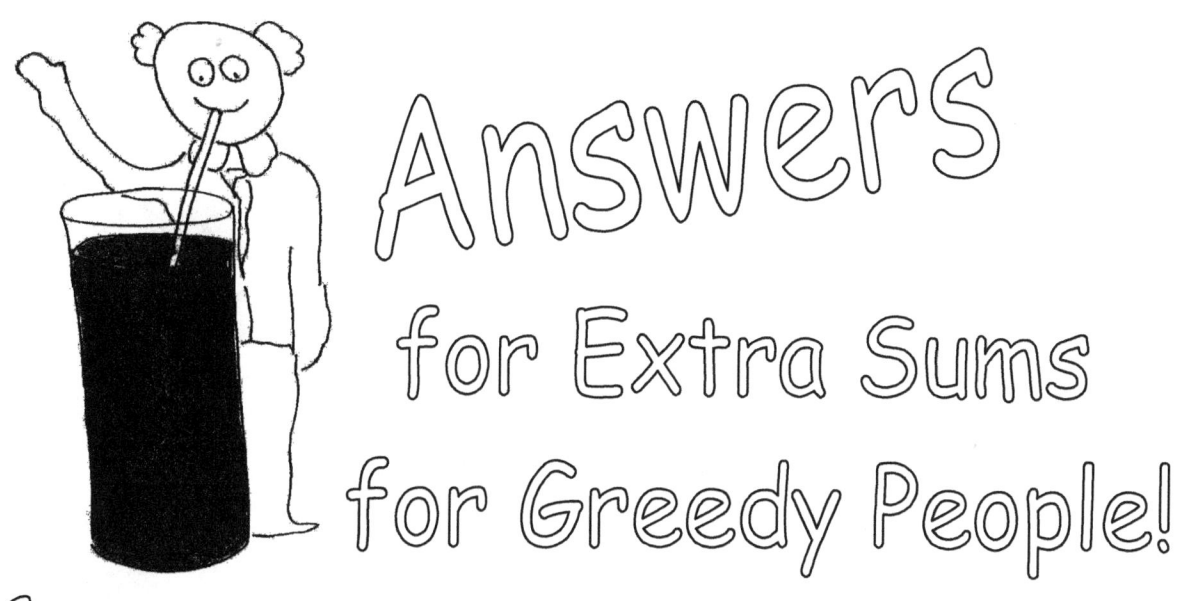

Answers for Extra Sums for Greedy People!

1 Well Done!
You have made the next page of Comic 123.

2
There are <u>4</u> pieces.
There are <u>9</u> pieces.
There are <u>16</u> pieces.
There are <u>25</u> pieces.

3

15

20

25

Here is a number face using 0 0 1 3 5 and 7	The answer is one that curves round then flicks back.
	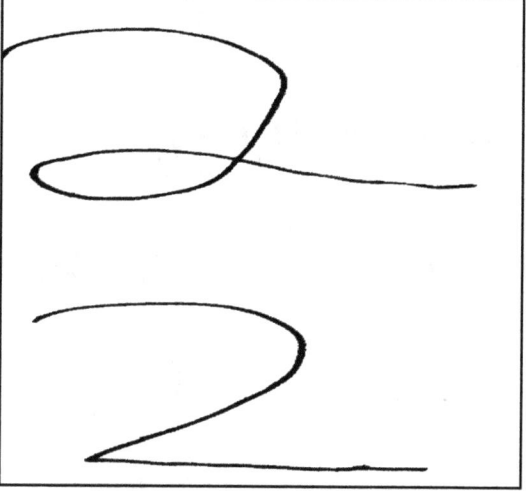

You could say ...

LOOK
You have <u>100</u> noses on your face and you are wearing <u>0</u> pairs of trousers.

LOOK
There are <u>0</u> windows in that house and it has <u>1000</u> chimney pots.

A B C D E F̲

A C E G I K̲

2 4 6 8 10 1̲2̲

5 10 15 20 25 3̲0̲

6 11 16 21 26 3̲1̲

10 20 30 40 50 6̲0̲

100 90 80 70 60 5̲0̲

M T W T F S̲

(First letter of days of the week)

J F M A M J̲

(First letter of months of the year)

1 2 3 1 2 3 1 2̲

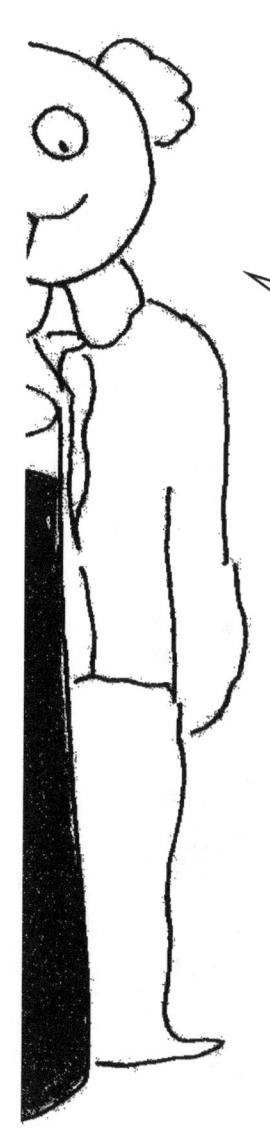

7

Have fun!
That's the answer!

8

Don't be
late asking
the time!

9 You should win.
Cardboard boxes can't fight!

10 Have a good party!

Oh no! 8 of my pigtails are missing in this book.

Coming Soon ...

```
    2
+   1
─────
    3
```

```
    3  2
+   2  1
─────────
    5  3
```

```
    4  3  2
+   3  2  1
────────────
    7  5  3
```

Mistakes

Some of Anne's pigtails are missing. How many?

(See page 108 for the answer.)

Brian has done his best not to make mistakes and is sorry for any he has not seen.

(Apart from Anne's missing pigtails that is!)

COMIC MATHS: Sue; fantasy-based learning for 4, 5 and 6 year olds
ISBN 978-0-9561602-1-8

First Edition

© Brian Williamson 2012

Cover design and Sue on pages 3 and 8 by Kathryn Wilson, character concepts by Brian Williamson. Illustrations in the text by Brian Williamson apart from: - Comic Maths word art and images on pages 3, 6, 8, 109 and 133 by Kathryn Wilson, John's 1 and 2 in comic number 4 by George Williamson, improved larger footballs by Alice Williamson. Copy editing by George and Brian Williamson.

The right of Brian Williamson to be identified as the author and illustrator of this work has been asserted in accordance with Sections 77 and 78 of the Copyright, Designs and Patents Act 1988.

First published in 2012 The Captain Papadopoulos Publishing Company

Distributed by The Comic Maths Community Interest Company, www.ComicMaths.co.uk

All rights reserved. No part of this publication may be reproduced, stored in a retrieval system or transmitted, in any form or by any means, electronic, mechanical, photocopying, recording or otherwise, without the prior permission of the author.

A Cataloging-in-Publication (CIP) record of this book is held at the British Library and at the Library of Congress

www.ingramcontent.com/pod-product-compliance
Lightning Source LLC
Chambersburg PA
CBHW080523030426
42337CB00023B/4606